Published 2012 by Cliff Top Press Ltd.
www.arithmeticvillage.com

ISBN 978-0-9845731-7-2

Printed by Lightning Source, USA

Files licensed by www.depositphotos.com: Leather © montego • Background of burlap hessian sacking © odua
• Cardboard background © Leonardi

Linus Minus

By Kimberly Moore

Linus Minus, an adorable mess
starts with more jewels
and ends up with less.

With fifteen jewels he runs out to play,

leaping along, loses three on the way.

With the twelve jewels, he checks out some waves.

Two treasures he drops,

the others he saves.

He fishes with ten and throws out a line.

A fish eats one gem, leaving Linus nine.

He crosses a farm and chases some chicks,

scattering three jewels, until he keeps six.

With the six jewels, Linus climbs a tree.

When he is finished,
he only has three.

Playing in the brook always having fun,

two more splash down,
he gives the King one.

His worth in jewels is never measured.

Everyone here is equally treasured.

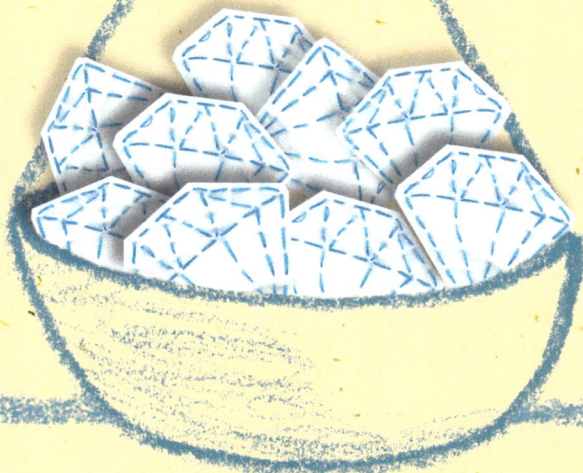

To mimic Linus, you know how to act,

just lose some jewels, and you will subtract!

Dear Guardian,

Linus Minus simply demonstrates the concept of subtraction.

Some children may want to act like Linus and lose objects!
Have a contest to see who can lose the most items.

As number familiarity increases, keep track of how much you
lose or write down the numerical equations for each page.

With a gentle, light hearted approach, these experiences can
support a lifetime love of math!

For more inspirational ideas visit www.arithmeticvillage.com

Lost without you,

Village Teachers Association

Kimberly

Arithmetic Village

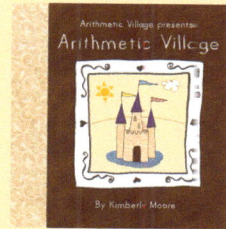

Polly Plus

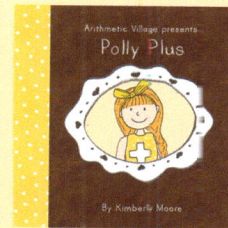

Linus Minus

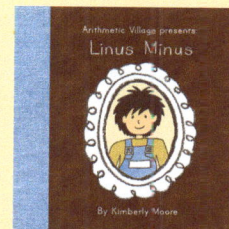

Tina Times

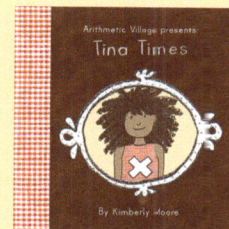

King David Divide

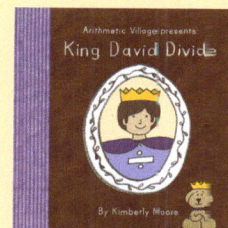

www.ingramcontent.com/pod-product-compliance
Lightning Source LLC
Chambersburg PA
CBHW040026050426
42452CB00003B/148